Fibonacci Numbers

Fibonacci Numbers

Nikolai Nikolaevich Vorob'ev

Translated from the Russian by
Halina Moss

Translation Edited by
Ian N. Sneddon

Dover Publications, Inc.
Mineola, New York

Bibliographical Note

This Dover edition, first published in 2011, is an unabridged republication of the work originally published in 1961 by Pergamon Press Ltd., Oxford.

Library of Congress Cataloging-in-Publication Data

Vorob'ev N. N. (Nikolai Nikolaevich), 1925–
  [Chisla Fibonachchi. English]
  Fibonacci numbers / Nikolai Nikolaevich Vorob'ev. — Dover ed.
    p. cm.
  Originally published: Oxford ; New York : Pergamon Press, 1961.
  ISBN-13: 978-0-486-48386-3
  ISBN-10: 0-486-48386-X
  1. Fibonacci numbers. I. Title.

QA241.V613 2011
512.7'2—dc23

2011016642

Manufactured in the United States
48386X03   2019
www.doverpublications.com

CONTENTS

v

Foreword

In elementary mathematics there are many difficult and interesting problems not connected with the name of an individual, but rather possessing the character of a kind of "mathematical folklore". Such problems are scattered throughout the wide literature of popular (or, simply, entertaining!) mathematics, and often it is very difficult to establish the source of a particular problem.

These problems often circulate in several versions. Sometimes several such problems combine into a single, more complex, one, sometimes the opposite happens and one problem splits up into several simple ones: thus it is often difficult to distinguish between the end of one problem and the beginning of another. We should consider that in each of these problems we are dealing with little mathematical theories, each with its own history, its own complex of problems and its own characteristic methods, all, however, closely connected with the history and methods of "great mathematics".

The theory of Fibonacci numbers is just such a theory. Derived from the famous "rabbit problem", going back nearly 750 years, Fibonacci numbers, even now, provide one of the most fascinating chapters of elementary mathematics. Problems connected with Fibonacci numbers occur in many popular books on mathematics, are discussed at meetings of school mathematical societies, and feature in mathematical competitions.

The present booklet contains a set of problems which were the themes of several meetings of the schoolchildren's mathematical club of Leningrad State University in

the academic year 1949-50. In accordance with the wishes
of those taking part, the questions discussed at these
meetings were mostly number-theoretical, a theme which is
developed in greater detail here.

This book is designed to appeal basically to pupils of
16 or 17 years of age in a high school. The concept of a
limit is met with only in examples 7 and 8 in chapter
III. The reader who is not acquainted with this concept
can omit these without prejudice to his understanding of
what follows. That applies also to binomial coefficients
(I, example 8) and to trigonometry (IV, examples 2 & 3).
The elements which are presented of the theory of divisibility and of the theory of continued fractions do not
presuppose any knowledge beyond the limits of a school
course.

Those readers who develop an interest in the principle
of constructing recurrent series are recommended to read
the small but full booklet of A. I. Markushevich, "Recurrent Sequences" (Vozvratnyye posledovatel'nosti)
(Gostekhizdat, 1950). Those who become interested in
facts relating to the theory of numbers are referred to
textbooks in this subject*.

---

* English-speaking readers are referred to
H. Davenport, "The Higher Arithmetic" (London, Hutchinson,
1952);
Burton W. Jones, "The Theory of Numbers" (London, Constable,
1955).

INTRODUCTION

1. The ancient world was rich in outstanding mathematicians. Many achievements of ancient mathematics are admired to this day for the acuteness of mind of their authors, and the names of Euclid, Archimedes and Hero are known to every educated person.

Things are different as far as the mathematics of the Middle Ages is concerned. Apart from Vieta, who lived as late as the sixteenth century, and mathematicians closer in time to us, a school course of mathematics does not mention a single name connected with the Middle Ages. This is, of course, no accident. In that epoch the science developed extremely slowly, and mathematicians of real stature were few.

The greater then is the interest of the work *Liber Abacci* ("a book about the abacus"), written by the remarkable Italian mathematician, Leonardo of Pisa, who is better known by his nickname Fibonacci (an abbreviation of *filius Bonacci*). This book, written in 1202, has survived in its second version, belonging to 1228.

Liber Abacci is a voluminous work, containing nearly all the arithmetical and algebraic knowledge of those times. It played a notable part in the development of mathematics in Western Europe in subsequent centuries. In particular, it was from this book that Europeans became acquainted with the Hindu (Arabic) numerals.

The theory contained in *Liber Abacci* is illustrated by

1

a great many examples, which make up a significant part
of the book.

Let us consider one of these examples, that which can
be found on pages 123-124 of the manuscript of 1228:

"How many pairs of rabbits are born of one pair in a year?"

This problem is stated in the form:-

"Someone placed a pair of rabbits in a certain place,
enclosed on all sides by a wall, to find out how many
pairs of rabbits will be born there in the course of one
year, it being assumed that every month a pair of rabbits
produces another pair, and that rabbits begin to bear
young two months after their own birth.

"As the first pair produces issue in the first month,
in this month there will be 2 pairs. Of these, one pair,
namely the first one, gives birth in the following month,
so that in the second month there will be 3 pairs. Of
these, 2 pairs will produce issue in the following month,
so that in the third month 2 more pairs of rabbits will
be born, and the number of pairs of rabbits in that month
will reach 5; of which 3 pairs will produce issue in
the fourth month, so that the number of pairs of rabbits
will then reach 8. Of these, 5 pairs will produce a
further 5 pairs, which, added to the 8 pairs, will give 13
pairs in the fifth month. Of these, 5 pairs do not
produce issue in that month but the other 8 do, so that
in the sixth month 21 pairs result. Adding the 13 pairs
that will be born in the seventh month, 34 pairs are
obtained: added to the 21 pairs born in the eighth month
it becomes 55 pairs in that month: this, added to the 34
pairs born in the ninth month, becomes 89 pairs: and in-
creased again by 55 pairs which are born in the tenth
month, makes 144 pairs in that month. Adding the 89
further pairs which are born in the eleventh month, we
get 233 pairs, to which we add, lastly, the 144 pairs
born in the final month. We thus obtain 377 pairs: this

is the number of pairs procreated from the first pair by
the end of one year.

```
┌─────────────────────────────┐
│          A pair             │
│            1                │
│      First   (Month)        │
│            2                │
│          Second             │
│            3                │
│          Third              │
│            5                │
│          Fourth             │
│            8                │
│          Fifth              │
│           13                │
│          Sixth              │
│           21                │
│          Seventh            │
│           34                │
│          Eighth             │
│           55                │
│          Ninth              │
│           89                │
│          Tenth              │
│           144               │
│          Eleventh           │
│           233               │
│          Twelfth            │
│           377               │
└─────────────────────────────┘
```

Fig. 1.

"From [Fig.1] * we see how we arrive at it: we add to
the first number the second one i.e. 1 and 2; the second
one to the third; the third to the fourth; the fourth
to the fifth; and in this way, one after another, until
we add together the tenth and the eleventh numbers (i.e.

---

* Fibonacci does all calculation tables and diagrams in the
  margin.

144 and 233) and obtain the total number of rabbits (i.e. 377); and it is possible to do this in this order for an infinite number of months".

2. We now pass from rabbits to numbers and examine the following numerical sequence

$$u_1, u_2, \ldots, u_n, \tag{1}$$

in which each term equals the sum of two preceding terms, i.e. for any $n \geqslant 2$

$$u_n = u_{n-1} + u_{n-2}. \tag{2}$$

Such sequences, in which each term is defined as some function of the previous ones, are met with often in mathematics, and are called *recurrent sequences*. The process of successive definition of the elements of such sequences is itself called the *recurrence process,* and equation (2) is called a *recurrence relation.* The reader can find the elements of the general theory of recurrent sequences in the book by Markushevich mentioned above.

We note that we cannot calculate the terms of sequence (1) by condition (2) above.

It is possible to make up any number of different numerical sequences satisfying this condition. For example

$$2, \ 5, \ 7, \ 12, \ 19, \ 31, \ 50, \ \ldots,$$
$$1, \ 3, \ 4, \ \ 7, \ 11, \ 18, \ 29, \ \ldots,$$
$$-1, \ -5, \ -6, \ -11, \ -17, \ \ldots \text{ and so on.}$$

This means that for the unique construction of sequence (1) the condition (2) is obviously inadequate, and we must establish certain supplementary conditions. For example, we can fix the first few terms of sequence (1).

How many of the first terms of sequence (1) must we fix
so that it is possible to calculate all its following
terms, using only condition (2)?

We begin by pointing out that not every term of sequence (1) can be obtained by (2) if only because not all
terms of (1) have two preceding ones;   for instance, the
first term of the sequence has no terms preceding it, and
the second term is preceded by only one.   This means that
in addition to condition (2) we must know the first two
terms of the sequence in order to define it.

This is obviously sufficient to enable us to calculate
any term of sequence (1).   Indeed, u_3 can be calculated
as the sum of the prescribed u_1 and u_2; u_4 as the sum of
u_2 and the previously calculated u_3; u_5 as the sum of
the previously calculated u_3 and u_4 and so on "in this
order up to an infinite number of terms".

Passing thus from two neighbouring terms to the one
immediately following them, we can reach the term with
any required suffix and calculate it.

<u>3.</u>   Let us now turn to the important particular case of
sequence (1), where $u_1 = 1$ and $u_2 = 1$.   As was pointed
out above, condition (2) enables us to calculate succes-
sively the terms of this series.   It is easy to verify
that in this case the first 13 terms are the numbers

1, 1, 2, 3, 5, 8, 13, 21, 34, 55, 89, 144, 233, 377,

which we already met in the rabbit problem.   To honour
the author of the problem, sequence (1) when $u_1 = u_2 = 1$
is called the *Fibonacci sequence,* and its terms are known
as *Fibonacci numbers.*

Fibonacci numbers possess a number of interesting and
important properties, which are the subject of this whole
booklet.

I

THE SIMPLEST PROPERTIES OF FIBONACCI NUMBERS

1. To begin with we shall calculate the sum of the first n Fibonacci numbers. We shall show that

$$u_1 + u_2 + \cdots + u_n = u_{n+2} - 1. \tag{3}$$

Indeed, we have:

$$u_1 = u_3 - u_2,$$

$$u_2 = u_4 - u_3,$$

$$u_3 = u_5 - u_4,$$

$$\cdots \cdots \cdots$$

$$u_{n-1} = u_{n+1} - u_n,$$

$$u_n = u_{n+2} - u_{n+1}.$$

Adding up these equations term by term we obtain

$$u_1 + u_2 + \cdots + u_n = u_{n+2} - u_2,$$

and all that remains is to remember that $u_2 = 1$.

2. The sum of Fibonacci numbers with odd suffixes

$$u_1 + u_3 + u_5 + \ldots + u_{2n-1} = u_{2n}. \qquad (4)$$

To establish this equation we shall write

$$u_1 = u_2,$$

$$u_3 = u_4 - u_2,$$

$$u_5 = u_6 - u_4,$$

$$\cdots \cdots \cdots$$

$$u_{2n-1} = u_{2n} - u_{2n-2}.$$

Adding these equations term by term we obtain the required result.

3. The sum of Fibonacci numbers with even suffixes

$$u_2 + u_4 + \ldots + u_{2n} = u_{2n+1} - 1. \qquad (5)$$

From section 1 we have

$$u_1 + u_2 + u_3 + \ldots + u_{2n} = u_{2n+2} - 1;$$

subtracting equation (4) from the above equation we obtain

$$u_2 + u_4 + \ldots + u_{2n} = u_{2n+2} - 1 - u_{2n} = u_{2n+1} - 1,$$

as was required.

Further, subtracting (5) from (4) term by term we get

$$u_1 - u_2 + u_3 - u_4 + \ldots + u_{2n-1} - u_{2n} = -u_{2n-1} + 1.$$

$$(6)$$

Now, let us add u_{2n+1} to both sides of (6)

$$u_1 - u_2 + u_3 - u_4 + \cdots - u_{2n} + u_{2n+1} = u_{2n} + 1. \quad (7)$$

Combining (6) and (7) we get for the sum of Fibonacci numbers with alternating signs:

$$u_1 - u_2 + u_3 - u_4 + \cdots + (-1)^{n+1} u_n =$$
$$= (-1)^{n+1} u_{n-1} + 1. \quad (8)$$

4. The formulae (3) and (4) were deduced by means of the term by term addition of a whole series of obvious equations. A further example of the application of this procedure is the proof of the formula for the sum of squares of the first n Fibonacci numbers

$$u_1^2 + u_2^2 + \cdots + u_n^2 = u_n u_{n+1}. \quad (9)$$

We note that

$$u_k u_{k+1} - u_{k-1} u_k = u_k (u_{k+1} - u_{k-1}) = u_k^2.$$

Adding up the equations

$$u_1^2 = u_1 u_2,$$

$$u_2^2 = u_2 u_3 - u_1 u_2,$$

$$u_3^2 = u_3 u_4 - u_2 u_3,$$

$$\cdots \cdots \cdots \cdots$$

$$u_n^2 = u_n u_{n+1} - u_{n-1} u_n$$

term by term, we obtain (9).

5. Many relationships between Fibonacci numbers are conveniently proved with the aid of the method of induction.

The essence of the method of induction is as follows. In order to prove that a certain proposition is correct for any natural number it is sufficient to establish:

(a) that it holds for the number 1;

(b) that from the truth of the proposition for an arbitrary natural number n follows its truth for the number $n + 1$.

Any inductive proof of a proposition true for any natural number consists, therefore, of two parts.

In the first part the truth of the proposition being proved is established for $n = 1$. The truth of the proposition for $n = 1$ is sometimes called the *basis of induction*.

In the second part of the proof the truth of the proposition is assumed for a certain arbitrary (but fixed) number n, and from this assumption, often called the inductive assumption, the deduction is made that the proposition is also true for the number $n + 1$. The second part of the proof is called the *inductive transition*.

The detailed presentation of the method of induction and numerous examples of the application of different forms of this method can be found in I.S. Sominskii, "The Method of Mathematical Induction".* Thus, in particular, the version of the method of induction with the inductive transition "from n and $n + 1$ to $n + 2$" employed by us below is given in Sominskii's book on page 9* and is illustrated there on page 16* by problems 18 and 19.

---

* English edition, Pergamon Press, 1961.

We prove by induction the following important formula:

$$u_{n+m} = u_{n-1}u_m + u_n u_{m+1}. \tag{10}$$

We shall carry out the proof of this formula by induction on m. For $m = 1$ this formula takes the form

$$u_{n+1} = u_{n-1}u_1 + u_n u_2 = u_{n-1} + u_n,$$

which is obviously true. For $m = 2$ formula (10) is also true, because

$$u_{n+2} = u_{n-1}u_2 + u_n u_3 = u_{n-1} + 2u_n =$$
$$= u_{n-1} + u_n + u_n = u_{n+1} + u_n.$$

Thus the basis of the induction is proved. The inductive transition can be proved in this form: supposing formula (10) to be true for $m = k$ and for $m = k + 1$, we shall prove that it also holds when $m = k + 2$.

Thus, let

$$u_{n+k} = u_{n-1}u_k + u_n u_{k+1}$$

and

$$u_{n+k+1} = u_{n-1}u_{k+1} + u_n u_{k+2}.$$

Adding the last two equations term by term we obtain

$$u_{n+k+2} = u_{n-1}u_{k+2} + u_n u_{k+3},$$

and this was the required result.

Putting $m = n$ in formula (10) we obtain

$$u_{2n} = u_{n-1}u_n + u_n u_{n+1},$$

or

$$u_{2n} = u_n(u_{n-1} + u_{n+1}).\qquad(11)$$

From this last equation it is obvious that u_{2n} is divisible by u_n. In the next chapter we shall prove a much more general result.

Since

$$u_n = u_{n+1} - u_{n-1},$$

formula (11) can be rewritten thus:

$$u_{2n} = (u_{n+1} - u_{n-1})(u_{n+1} + u_{n-1}),$$

or

$$u_{2n} = u_{n+1}^2 - u_{n-1}^2,$$

i.e., the difference of the squares of two Fibonacci numbers whose positions in the sequence differ by two is again a Fibonacci number.

Similarly (taking $m = 2n$) it can be shown, that

$$u_{3n} = u_{n+1}^3 + u_n^3 - u_{n-1}^3.$$

6. The following formula will be found useful in what follows:

$$u_{n+1}^2 = u_n u_{n+2} + (-1)^n.\qquad(12)$$

Let us prove it by induction over n. For $n = 1$, (12) takes the form

$$u_2^2 = u_1 u_3 - 1,$$

which is obvious.

We now suppose formula (12) proved for a certain n. Adding $u_{n+1} u_{n+2}$ to both sides of it we obtain

$$u_{n+1}^2 + u_{n+1} u_{n+2} = u_n u_{n+2} + u_{n+1} u_{n+2} + (-1)^n$$

or

$$u_{n+1}(u_{n+1} + u_{n+2}) = u_{n+2}(u_n + u_{n+1}) + (-1)^n,$$

or

$$u_{n+1} u_{n+3} = u_{n+2}^2 + (-1)^n,$$

or

$$u_{n+2}^2 = u_{n+1} u_{n+3} + (-1)^{n+1}$$

Thus, the inductive transition is established and formula (12) is proved for any n.

<u>7.</u>   In a similar way, it is possible to establish the following properties of Fibonacci numbers:

$$u_1 u_2 + u_2 u_3 + u_3 u_4 + \ldots + u_{2n-1} u_{2n} = u_{2n}^2,$$

$$u_1 u_2 + u_2 u_3 + u_3 u_4 + \ldots + u_{2n} u_{2n+1} = u_{2n+1}^2 - 1,$$

$$n u_1 + (n-1) u_2 + (n-2) u_3 + \ldots + 2 u_{n-1} + u_n =$$
$$= u_{n+4} - (n+3).$$

The proofs are left to the reader.

8. It turns out that there is a connection between the
Fibonacci numbers and another set of remarkable numbers – the
binomial coefficients. Let us set out the binomial coefficients* in the following triangle, called Pascal's triangle:

$$C_0^0$$

$$C_1^0 \quad C_1^1$$

$$C_2^0 \quad C_2^1 \quad C_2^2$$

$$C_3^0 \quad C_3^1 \quad C_3^2 \quad C_3^3$$

. . . .

i.e.

```
1

1   1

1   2   1

1   3   3   1

1   4   6   4   1

1   5   10  10  5   1

1   6   15  20  15  6   1
```

. . . .

The straight lines drawn through the numbers of this triangle
at an angle of 45 degrees to the rows we shall call "the rising
diagonals" of Pascal's triangle. For instance, the straight
lines passing through numbers 1, 4, 3, or 1, 5, 6, 1, are
rising diagonals.

We shall show that the sum of numbers lying along a certain
rising diagonal is a Fibonacci number.

Indeed, the first and topmost rising diagonal of Pascal's
triangle is merely 1, the first Fibonacci number. The second
diagonal also consists of 1. To prove the general proposition,

* Expressions of the form C_b^a, as used here and below in con-
formity with the original, represent the ^bC_a of customary
English usage.

it is sufficient to show that the sum of all numbers making up the $(n-2)$th and the $(n-1)$th diagonal of Pascal's triangle is equal to the sum of the numbers making up the nth diagonal.

On the $(n-2)$th diagonal we have the numbers

$$C_{n-3}^0, \ C_{n-4}^1, \ C_{n-5}^2, \ \dots,$$

and on the $(n-1)$th diagonal the numbers

$$C_{n-2}^0, \ C_{n-3}^1, \ C_{n-4}^2, \ \dots$$

The sum of all these numbers can be written thus

$$C_{n-2}^0 + (C_{n-3}^0 + C_{n-3}^1) + (C_{n-4}^1 + C_{n-4}^2) + \dots \qquad (13)$$

But for binomial coefficients

$$C_{n-2}^0 = C_{n-1}^0 = 1$$

and

$$C_k^i + C_k^{i+1} = \frac{k(k-1) \ \dots \ (k-i+1)}{1.2. \ \dots \ .i} +$$

$$+ \frac{k(k-1) \ \dots \ (k-i+1)(k-i)}{1.2. \ \dots \ .i.(i+1)} =$$

$$= \frac{k(k-1) \ \dots \ (k-i+1)}{1.2. \ \dots \ .i} \left(1 + \frac{k-i}{i+1}\right) =$$

$$= \frac{k(k-1) \ \dots \ (k-i+1)}{1.2. \ \dots \ .i} \cdot \frac{i+1+k-i}{i+1} =$$

$$= \frac{(k+1)k(k-1) \ \dots \ (k-i+1)}{1.2. \ \dots \ .i.(i+1)} = C_{k+1}^{i+1}.$$

Expression (13) therefore equals

$$C_{n-1}^0 + C_{n-2}^1 + C_{n-3}^2 + \cdots,$$

i.e. the sum of the numbers lying on the nth diagonal of the triangle.

From this proof and formula (3) we immediately get: The sum of all binomial coefficients lying above the nth rising diagonal of Pascal's triangle (inclusive of that diagonal) equals $u_{n+2} - 1$.

Making use of formulae (4), (5), (6) and similar ones, the reader can easily obtain further identities connecting Fibonacci numbers with binomial coefficients.

9. So far, we have defined Fibonacci numbers by a recurrence procedure, i.e. inductively, by their suffixes. It turns out, however, that any Fibonacci number can also be defined directly, as a function of its suffix.

To see this, we investigate various sequences satisfying the relationship (2). We shall call all such sequences solutions of equation (2).

In future we shall denote the sequences

$$v_1, \ v_2, \ v_3, \ \cdots,$$

$$v_1', \ v_2', \ v_3', \ \cdots,$$

$$v_1'', \ v_2'', \ v_3'', \ \cdots,$$

by V, V' and V'' respectively.

To begin with we shall prove two simple lemmas.

__Lemma 1.__ *If V is the solution of equation (2) and c is an arbitrary number, then the sequence cV (i.e. the sequence* cv_1, cv_2, cv_3, *...) is also a solution of equation (2).*

__Proof.__ Multiplying the relationship

$$v_n = v_{n-2} + v_{n-1}$$

term by term by c, we get

$$cv_n = cv_{n-2} + cv_{n-1},$$

as was required.

__Lemma 2.__ *If the sequences V' and V" are solutions of (2), then their sum V' + V" (i.e. the sequence* $v_1' + v_1"$, $v_2' + v_2"$, $v_3' + v_3"$, *...) is also a solution of (2).*

__Proof:__ From the conditions stated in the lemma we have

$$v_n' = v_{n-1}' + v_{n-2}'$$

and

$$v_n" = v_{n-1}" + v_{n-2}".$$

Adding these two equations term by term, we get

$$v_n' + v_n" = (v_{n-1}' + v_{n-1}") + (v_{n-2}' + v_{n-2}").$$

Thus, the lemma is proved.

Now, let V' and $V"$ be two solutions of equation (2) which are not proportional. We shall show that any sequence V which is a solution of equation (2) can be written in the form

$$c_1 V_1' + c_2 V_2', \tag{14}$$

where c_1 and c_2 are constants. It is therefore usual to speak of (14) as the general solution of the equation (2).

First of all, we shall prove that if solutions of (2) V' and V'' are not proportional, then

$$\frac{v_1'}{v_1''} \neq \frac{v_2'}{v_2''} . \tag{15}$$

The proof of (15) is carried out by assuming the opposite.

For solutions V' and V'' of (2) which are not proportional, let

$$\frac{v_1'}{v_1''} = \frac{v_2'}{v_2''} . \tag{16}$$

On writing down the derived proportion we get

$$\frac{v_1' + v_2'}{v_1'' + v_2''} = \frac{v_2'}{v_2''}$$

or, taking into account that V' and V'' are solutions of equation (2),

$$\frac{v_3'}{v_3''} = \frac{v_2'}{v_2''} .$$

Similarly, we convince ourselves (by induction!) that

$$\frac{v_3'}{v_3''} = \frac{v_4'}{v_4''} = \ldots = \frac{v_n'}{v_n''} = \ldots$$

Thus, it follows from (16) that the sequences V' and V'' are proportional, which contradicts the assumption. This means that (15) is true.

Now, let us take a certain sequence V, which is a solution of the equation (2). This sequence, as was pointed out in section 2 of the Introduction, is fully defined if its two first terms, v_1 and v_2, are given.

Let us find such c_1 and c_2, that

$$c_1 v'_1 + c_2 v''_1 = v_1,$$
$$c_1 v'_2 + c_2 v''_2 = v_2. \tag{17}$$

Then, on the basis of lemmas 1 and 2, $c_1 V' + c_2 V''$ gives us the sequence V.

In view of condition (15), the simultaneous equations (17) are soluble with respect to c_1 and c_2 no matter what the numbers v_1 and v_2 are:

$$c_1 = \frac{v_1 v''_2 - v_2 v''_1}{v'_1 v''_2 - v''_1 v'_2}, \qquad c_2 = \frac{v'_1 v_2 - v'_2 v_1}{v'_1 v''_2 - v''_1 v'_2}$$

[By the condition (15) the denominator does not equal zero].

Substituting the values of c_1 and c_2 thus calculated in (14) we obtain the required representation of the sequence V.

This means that in order to describe all solutions of equation (2) it is sufficient to find any two solutions of it which are not proportional.

Let us look for these solutions among geometric progressions. In accordance with lemma 1 it is sufficient to limit ourselves to the consideration of only those progressions whose first term is equal to unity. Thus, let us take the progression

$$1, \ q, \ q^2, \ \ldots$$

In order that this progression should be a solution of (2) it is necessary that for any n the equality

$$q^{n-2} + q^{n-1} = q^n$$

should be fulfilled. Or, dividing by q^{n-2},

$$1 + q = q^2.$$

The roots of this quadratic equation, i.e. $\dfrac{1 + \sqrt{5}}{2}$ and $\dfrac{1 - \sqrt{5}}{2}$, will be the required common ratios of the progressions. We shall denote them by α and β respectively. Note that $\alpha\beta = -1$.

We have thus obtained two geometric progressions which are solutions of (2). Therefore all sequences of the form

$$c_1 + c_2, \; c_1\alpha + c_2\beta, c_1\alpha^2 + c_2\beta^2, \; \ldots \qquad (18)$$

are solutions of (2). As the progressions found by us have different common ratios and are therefore not proportional, formula (18) gives us all solutions of equation (2).

In particular, for certain values of c_1 and c_2 formula (18) should give us also the Fibonacci series. For this, as was pointed out above, it is necessary to find c_1 and c_2 from the equations

$$c_1 + c_2 = u_1$$

and

$$c_1\alpha + c_2\beta = u_2,$$

i.e. from the simultaneous equations

$$c_1 + c_2 = 1,$$

$$c_1 \frac{1 + \sqrt{5}}{2} + c_2 \frac{1 - \sqrt{5}}{2} = 1.$$

Having solved them, we get

$$c_1 = \frac{1 + \sqrt{5}}{2\sqrt{5}}, \quad c_2 = -\frac{1 - \sqrt{5}}{2\sqrt{5}},$$

whence

$$u_n = c_1 a^{n-1} + c_2 \beta^{n-1} =$$

$$= \frac{1 + \sqrt{5}}{2\sqrt{5}}\left(\frac{1 + \sqrt{5}}{2}\right)^{n-1} - \frac{1 - \sqrt{5}}{2\sqrt{5}}\left(\frac{1 - \sqrt{5}}{2}\right)^{n-1},$$

i.e.

$$u_n = \frac{\left(\dfrac{1 + \sqrt{5}}{2}\right)^n - \left(\dfrac{1 - \sqrt{5}}{2}\right)^n}{\sqrt{5}} \tag{19}$$

Formula (19) is called *Binet's formula* in honour of the mathematician who first proved it. Obviously, similar formulae can be derived for other solutions of (2). The reader should do it for the sequences introduced in section 2 of the Introduction.

10. With the help of Binet's formula it is easy to find the sums of many series connected with Fibonacci numbers.

For instance, we can find the sum

$$u_3 + u_6 + u_9 + \ldots + u_{3n}.$$

We have

$$u_3 + u_6 + \ldots + u_{3n} = \frac{a^3 - \beta^3}{\sqrt{5}} + \frac{a^6 - \beta^6}{\sqrt{5}} + \ldots + \frac{a^{3n} - \beta^{3n}}{\sqrt{5}} =$$

$$= \frac{1}{\sqrt{5}}(\alpha^3 + \alpha^6 + \ldots + \alpha^{3n} - \beta^3 - \beta^6 - \ldots - \beta^{3n}),$$

or, having summed the geometric progressions involved,

$$u_3 + u_6 + \ldots + u_{3n} = \frac{1}{\sqrt{5}}\left(\frac{\alpha^{3n+3} - \alpha^3}{\alpha^3 - 1} - \frac{\beta^{3n+3} - \beta^3}{\beta^3 - 1}\right).$$

But

$$\alpha^3 - 1 = \alpha + \alpha^2 - 1 = \alpha + \alpha + 1 - 1 = 2\alpha,$$

and similarly $\beta^3 - 1 = 2\beta$. Therefore

$$u_3 + u_6 + \ldots + u_{3n} = \frac{1}{\sqrt{5}}\left(\frac{\alpha^{3n+3} - \alpha^3}{2\alpha} - \frac{\beta^{3n+3} - \beta^3}{2\beta}\right).$$

or after cancellations

$$u_3 + u_6 + \ldots + u_{3n} = \frac{1}{\sqrt{5}}\left(\frac{\alpha^{3n+2} - \alpha^2 - \beta^{3n+2} + \beta^2}{2}\right) =$$

$$= \frac{1}{2}\left(\frac{\alpha^{3n+2} - \beta^{3n+2}}{\sqrt{5}} - \frac{\alpha^2 - \beta^2}{\sqrt{5}}\right) =$$

$$= \frac{1}{2}(u_{3n+2} - u_2) = \frac{u_{3n+2} - 1}{2}.$$

11. As another example of the application of Binet's formula, we shall calculate the sum of the cubes of the first n Fibonacci numbers.

We note that

$$u_k^3 = \left(\frac{\alpha^k - \beta^k}{\sqrt{5}}\right)^3 = \frac{1}{5}\left(\frac{\alpha^{3k} - 3\alpha^{2k}\beta^k + 3\alpha^k\beta^{2k} - \beta^{3k}}{\sqrt{5}}\right) =$$

$$= \frac{1}{5}\left(\frac{\alpha^{3k} - \beta^{3k}}{\sqrt{5}} - 3\alpha^k \beta^k \; \frac{\alpha^k - \beta^k}{\sqrt{5}} \right) =$$

$$= \frac{1}{5}(u_{3k} - (-1)^k 3u_k) = \frac{1}{5}(u_{3k} + (-1)^{k+1} 3u_k).$$

Therefore:

$$u_1^3 + u_2^3 + \dots + u_n^3 =$$

$$= \frac{1}{5}\left[(u_3 + u_6 + \dots + u_{3n}) + 3(u_1 - u_2 + u_3 - \dots + (-1)^{n+1}u_n)\right],$$

or, using formula (8) and the results of the preceding section,

$$u_1^3 + u_2^3 + \dots + u_n^3 = \frac{1}{5}\left(\frac{u_{3n+2} - 1}{2} + 3\left[1 + (-1)^{n+1}u_{n-1}\right] \right) =$$

$$= \frac{u_{3n+2} + (-1)^{n+1}6u_{n-1} + 5}{10}.$$

12. It is relevant to ask the question: how quickly do Fibonacci numbers grow with increasing suffix? Binet's formula gives us a sufficiently full answer even to this question.

It is not hard to prove the following theorem.

Theorem: *The Fibonacci number u_n is the nearest whole number to the nth term a_n of the geometric progression whose first term is $\dfrac{\alpha}{\sqrt{5}}$ and whose common ratio equals α*

Proof: Obviously it is sufficient to establish that the absolute value of the difference between u_n and a_n is always less than $\dfrac{1}{2}$. But

$$\left| u_n - a_n \right| = \frac{a^n - \beta^n}{\sqrt{5}} - \frac{a^n}{\sqrt{5}} = \left| \frac{a^n - a^n - \beta^n}{\sqrt{5}} \right| = \frac{|\beta|^n}{\sqrt{5}}.$$

As $\beta = -0.618\ldots$, therefore $|\beta| < 1$, and that means that for any n, $|\beta|^n < 1$ and even more so (since $\sqrt{5} > 2$) $\frac{|\beta|}{\sqrt{5}} < \frac{1}{2}$. The theorem is proved.

The reader who is acquainted with the theory of limits will be able to show by slightly altering the proof of this theorem that

$$\lim_{n \to \infty} \left| u_n - a_n \right| = 0.$$

Using this theorem it is possible to calculate Fibonacci numbers by means of logarithmic tables.

For instance, let us calculate u_{14} (u_{14} is the answer to the problem of Fibonacci about the rabbits):

$$\sqrt{5} = 2.2361, \qquad \log \sqrt{5} = 0.34949;$$

$$a = \frac{1 + \sqrt{5}}{2} = 1.6180, \qquad \log a = 0.20898;$$

$$\log \frac{a^{14}}{\sqrt{5}} = 14 \times 0.20898 - 0.34949 = 2.5762,$$

$$\frac{a^{14}}{\sqrt{5}} = 376.9.$$

The nearest whole number to 376.9 is 377; this is u_{14}.

When calculating Fibonacci numbers of very large suffixes, we can no longer calculate all the figures of the number by means of available tables of logarithms; we can only indicate the first few figures of it, so that the calculation turns out to be approximate.

As an exercise, the reader should prove that in the

decimal system, u_n for $n \geqslant 17$ has no more than $\dfrac{n}{4}$ and no fewer than $\dfrac{n}{5}$ figures. And of how many figures will u_{1000} consist?

II

NUMBER-THEORETIC PROPERTIES OF
FIBONACCI NUMBERS

Before we continue the study of Fibonacci numbers, we shall remind the reader of some of the simplest facts from the theory of numbers.

<u>1.</u> First, we shall indicate the process of finding the greatest common divisor of numbers a and b.

Suppose we divide a by b with a quotient equal to q_0 and a remainder r_1. Obviously, $a = bq_0 + r_1$ and $0 \leqslant r_1 < b$. Note that if $a < b$, $q_0 = 0$.

Let us further divide b by r_1 and let us denote the quotient by q_1 and the remainder by r_2. Obviously $b = r_1 q_1 + r_2$, and $0 \leqslant r_2 < r_1$. Since $r_1 < b$, therefore $q_1 \neq 0$. Then, dividing r_1 by r_2, we shall find $q_2 \neq 0$ and r_3 such that $r_1 = q_2 r_2 + r_3$ and $0 \leqslant r_3 < r_2$. We proceed in this manner for as long as it is possible to continue the process.

Sooner or later our process must terminate, since all the positive whole numbers r_1, r_2, r_3, ... are different, and every one of them is smaller than b. That means that their number does not exceed b, and the process should terminate no later than at the bth step. But it can only terminate when a certain division proves to be carried out perfectly, i.e. the remainder turns out equal to zero and it will be impossible to divide anything by it.

The process thus described bears the name of *Euclidean*

Algorithm. As a result of its application we obtain the following sequence of equations

$$
\left.
\begin{aligned}
a &= bq_0 + r_1, \\
b &= r_1 q_1 + r_2, \\
r_1 &= r_2 q_2 + r_3, \\
&\cdots\cdots\cdots \\
r_{n-2} &= r_{n-1}q_{n-1} + r_n, \\
r_{n-1} &= r_n q_n.
\end{aligned}
\right\} \tag{20}
$$

Let us examine the last non-zero remainder r_n. Obviously r_{n-1} is divisible by r_n. Let us now take the last but one equation in (20). On its right-hand side both terms are divisible by r_n and therefore r_{n-2} is divisible by r_n. Similarly, we show step by step (induction!) that r_{n-3}, r_{n-4}, ... and finally a and b are divisible by r_n. Thus, r_n is a common divisor of a and b. Let us show that r_n is the greatest common divisor of a and b. In order to do this, it is sufficient to show that any common divisor of a and b will also divide r_n.

Let d be a certain common divisor of a and b. From the first equation of (20) we notice that r_1 should be divisible by d. But, in that case, on the basis of the second equation of (20), r_2 is divisible by d. Similarly (induction!) we prove that d "goes into" r_3, ..., r_{n-1} and, finally, r_n.

We have thus proved that the Euclidean algorithm when applied to the natural numbers a and b does lead really to their greatest common divisor. This greatest common divisor of the numbers a and b is denoted by (a, b).

As an example, let us find $(u_{20}, u_{15}) = (6765, 610)$:

$$6765 = 610 \times 11 + 55,$$

$$610 = 55 \times 11 + 5,$$

$$55 = 5 \times 11$$

Thus, $(u_{20}, u_{15}) = 5 = u_5$. The fact that the greatest common divisor of two Fibonacci numbers turned out to be again a Fibonacci number is not accidental. It will be shown later that that is always the case.

2. There is an analogy between Euclid's algorithm and a process in geometry whereby the common measure of two commensurable segments is found.

Indeed, let us examine two segments, one of length a, the other of length b. Let us subtract the second segment from the first as many times as it is possible (if $b > a$, obviously we cannot do it even once) and denote the length of the remainder by r_1. Obviously $r_1 < b$. Now, let us subtract from the segment of length b the segment of length r_1 as many times as possible, and let us denote the newly obtained remainder by r_2. Carrying on in this manner, we obtain a sequence of remainders whose lengths, evidently, decrease. Up to this point, the resemblance to Euclid's algorithm is complete.

Subsequently, however, an important difference of the geometrical process from Euclid's algorithm for natural numbers is revealed. The sequence of remainders obtained from the subtraction of segments might not terminate, as the process of such subtraction can turn out to be capable of being continued indefinitely. This will happen if the chosen segments are incommensurable.

From the considerations in section 1, it follows that two segments whose lengths can be expressed by whole numbers are always commensurable.

We now establish several simple properties of the

greatest common divisor of two numbers.

3. (a, bc) is divisible by (a, b). Indeed, b, and
therefore bc, is divisible by (a, b); a is divisible by
(a, b) for obvious reasons. This means, according to the
proofs in section 1, that (a, bc) is divisible by (a, b) also.

4. $(ac, bc) = (a, b)c$

Proof: Let the equations (20) describe the process of
finding (a, b). Multiplying each of these equations by c
throughout, we shall, as is easily verified, obtain a set
of equations corresponding to the Euclidean algorithm as
applied to the numbers ac and bc. The last non-zero remainder here will be equal to r_nc, i.e. $(a, b)c$.

5. If $(a, c) = 1$, then $(a, bc) = (a, b)$. Indeed, (a, bc)
divides (ab, bc), according to section 3. But

$$(ab, bc) = (a, c)b = 1 \times b = b$$

in view of section 4. Thus b is divisible by (a, bc).
On the other hand (a, bc) divides a. By section 1 this
means that (a, bc) divides (a, b) also. And since according to section 3 (a, b) divides (a, bc) as well, then
$(a, b) = (a, bc)$.

6. a is divisible by b only if $(a, b) = b$. This is
obvious.

7. If c is divisible by b, then $(a, b) = (a + c, b)$.

Proof: Suppose that the application of the Euclidean
algorithm to the numbers a and b leads to the set of
equations (20). Let us apply the algorithm to the

numbers $a + c$ and b. Since c is divisible by b, as given, we can put $c = c_1 b$. The first step of the algorithm gives us the equation

$$a + c = (q_0 + c_1)b + r_1.$$

The subsequent steps of this algorithm will give us consecutively the second, third, etc., equations of the set (20). The last non-zero remainder is still r_n, and this means that $(a, b) = (a + c, b)$.

A useful exercise for the reader would be to prove this theorem on the sole basis of the results of sections 3-6, i.e. without a repeated reference to the idea of the Euclidean algorithm and to the set (20).

We now consider certain properties of Fibonacci numbers concerning their divisibility.

<u>8</u>. <u>Theorem</u>: *If n is divisible by m, then u_n is also divisible by u_m.*

<u>Proof</u>: Let n be divisible by m, i.e. let $n = mm_1$. We shall carry out the proof by induction over m_1. For $m_1 = 1$, $n = m$, so that in this case it is obvious that u_n is divisible by u_m. We now suppose that u_{mm_1} is divisible by u_m and consider $u_{m(m_1+1)}$. But $u_{m(m_1+1)} = u_{mm_1 + m}$ and, according to (10),

$$u_{m(m_1+1)} = u_{mm_1-1} u_m + u_{mm_1} u_{m+1}.$$

The first term of the right-hand side of this equation is obviously divisible by u_m. The second term contains u_{mm_1} as a factor, i.e. is divisible by u_m according to the inductive assumption. Hence, their sum, i.e. $u_{m(m_1+1)}$, is divisible by u_m as well. Thus the theorem is proved.

<u>9</u>. The topic of the arithmetical nature of Fibonacci

numbers (i.e. the nature of their divisors) is of great interest.

We prove that for a compound n other than 4, u_n is a compound number.

Indeed, for such an n we can write $n = n_1 n_2$, where $1 < n_1 < n$, $1 < n_2 < n$ and either $n_1 > 2$ or $n_2 > 2$. To be definite let $n_1 > 2$. Then, according to the theorem just proved, u_n is divisible by u_{n_1}, while $1 < u_{n_1} < u_n$ and this means that u_n is a compound number.

10. __Theorem.__ *Neighbouring Fibonacci numbers are prime to each other.*

__Proof:__ Let u_n and u_{n+1} have a certain common divisor $d > 1$, in contradiction of what the theorem states. Then their difference $u_{n+1} - u_n$ should be divisible by d. And since $u_{n+1} - u_n = u_{n-1}$, then u_{n-1} should be divisible by d. Similarly, we prove (induction!) that u_{n-2}, u_{n-3}, etc., and finally u_1, will be divisible by d. But $u_1 = 1$ therefore it cannot be divided by $d > 1$. The incompatibility thus obtained proves the theorem.

11. __Theorem.__ *For any m, n $(u_m, u_n) = u_{(m, n)}$.*

__Proof:__ To be definite, we suppose $m > n$, and apply the Euclidean algorithm to the numbers m and n:

$$m = n q_0 + r_1, \text{ where } 0 < r_1 < n,$$

$$n = r_1 q_1 + r_2, \text{ where } 0 < r_2 < r_1,$$

$$r_1 = r_2 q_2 + r_3, \text{ where } 0 < r_3 < r_2,$$

$$\cdots \cdots \cdots \cdots \cdots$$

$$r_{t-2} = r_{t-1} q_{t-1} + r_t, \text{ where } 0 < r_t < r_{t-1},$$

$$r_{t-1} = r_t q_t.$$

As we already know, r_t is the greatest common divisor of m and n.

Thus, $m = nq_0 + r_1$; this means that

$$(u_m, u_n) = (u_{nq_0+r_1}, u_n),$$

or, by equation (10),

$$(u_m, u_n) = (u_{nq_0-1}u_{r_1} + u_{nq_0}u_{r_1+1}, u_n),$$

or by sections 7, 8

$$(u_m, u_n) = (u_{nq_0-1}u_{r_1}, u_n),$$

or by sections 10, 5

$$(u_m, u_n) = (u_{r_1}, u_n).$$

Similarly, we prove that

$$(u_{r_1}, u_n) = (u_{r_2}, u_{r_1}),$$

$$(u_{r_2}, u_{r_1}) = (u_{r_3}, u_{r_2}),$$

$$\cdot \quad \cdot \quad \cdot \quad \cdot \quad \cdot \quad \cdot \quad \cdot \quad \cdot \quad \cdot$$

$$(u_{r_{t-1}}, u_{r_{t-2}}) = (u_{r_t}, u_{r_{t-1}}).$$

Combining all these equations, we get

$$(u_m, u_n) = (u_{r_t}, u_{r_{t-1}}),$$

and since r_{t-1} is divisible by r_t, $u_{r_{t-1}}$ is also divisible by u_{r_t}. Therefore $(u_{r_t}, u_{r_{t-1}}) = u_{r_t}$. Noting, finally, that $r_t = (m, n)$, we obtain the required result.

In particular, from the above proof we have a converse of the theorem in section 8: if u_n is divisible by u_m, then n is divisible by m. For if u_n is in fact divisible by u_m, then, according to section 6,

$$(u_n, u_m) = u_m. \qquad (21)$$

But we have proved that

$$(u_n, u_m) = u_{(n, m)}. \qquad (22)$$

Combining (21) and (22) we get

$$u_m = u_{(n, m)},$$

i.e. $m = (n, m)$, which means that n is divisible by m.

12. Combining the theorem in section 8 and the corollary to the theorem in section 11 we have: u_n is divisible by u_m if, and only if, n is divisible by m.

In view of this, the divisibility of Fibonacci numbers can be studied by studying the divisibility of their suffixes.

Let us find, for instance, some "signs of divisibility" of Fibonacci numbers. By "sign of divisibility" we mean a sign to show whether any particular Fibonacci number is divisible by a certain given number.

A Fibonacci number is even if, and only if, its suffix is divisible by 3.

A Fibonacci number is divisible by 3 if, and only if, its suffix is divisible by 4.

A Fibonacci number is divisible by 4 if, and only

if, its suffix is divisible by 6.

A Fibonacci number is divisible by 5 if, and only if, its suffix is divisible by 5.

A Fibonacci number is divisible by 7 if, and only if, its suffix is divisible by 8.

The proofs of all these signs of divisibility and all similar ones can be carried out easily by the reader, with the help of the proposition put forward at the beginning of the section, and by considering the third, fourth, sixth, fifth, eighth, etc. Fibonacci numbers respectively.

At the same time, the reader should prove that no Fibonacci number exists that would give a remainder of 4 when divided by 8; also, that there are no odd Fibonacci numbers divisible by 17.

13. Let us now take a certain whole number m. If there exists even one Fibonacci number u_n divisible by m, it is possible to find as many such Fibonacci numbers as desired. For example, such will be the numbers u_{2n}, u_{3n}, u_{4n}, ...

It would therefore be interesting to discover whether it is possible to find at least one Fibonacci number divisible by a given number m. It turns out that this is possible.

Let $\bar{k}$ be the remainder of the division of k by m, and let us write down a sequence of pairs of such remainders:

$$\langle \bar{u}_1, \bar{u}_2 \rangle, \langle \bar{u}_2, \bar{u}_3 \rangle, \langle \bar{u}_3, \bar{u}_4 \rangle, \ldots, \langle \bar{u}_n, \bar{u}_{n+1} \rangle, \ldots \tag{23}$$

If we regard pairs $\langle a_1, b_1 \rangle$ and $\langle a_2, b_2 \rangle$ as equal when $a_1 = a_2$ and $b_1 = b_2$, the number of different pairs of remainders of division by m equals m^2. If, therefore, we take the first $m^2 + 1$ terms of the sequence (23) there must be equal ones among them.

Let $\langle \bar{u}_k, \bar{u}_{k+1} \rangle$ be the first pair that repeats itself in the sequence (23). We shall show that this pair is $\langle 1, 1 \rangle$. Indeed, let us suppose the opposite, i.e. that the first repeated pair is the pair $\langle \bar{u}_k, \bar{u}_{k+1} \rangle$, where $k > 1$. Let us find in (23) a pair $\langle \bar{u}_l, \bar{u}_{l+1} \rangle$ $(l > k)$ equal to the pair $\langle \bar{u}_k, \bar{u}_{k+1} \rangle$. Since $u_{l-1} = u_{l+1} - u_l$ and $u_{k-1} = u_{k+1} - u_k$, and $\bar{u}_{l+1} = \bar{u}_{k+1}$ and $\bar{u}_l = \bar{u}_k$, the remainders of division of u_{l-1} and u_{k-1} by m are equal, i.e. $\bar{u}_{l-1} = \bar{u}_{k-1}$. However, it also follows that $\langle \bar{u}_{k-1}, \bar{u}_k \rangle = \langle \bar{u}_{l-1}, \bar{u}_l \rangle$, but the pair $\langle \bar{u}_{k-1}, \vec{u}_k \rangle$ is situated in the sequence (23) earlier than $\langle \bar{u}_k, \bar{u}_{k+1} \rangle$ and therefore $\langle \bar{u}_k, \bar{u}_{k+1} \rangle$ is not the first pair that repeated itself, which contradicts our premise. This means that the supposition $k > 1$ is wrong, and therefore $k = 1$.

Thus $\langle 1, 1 \rangle$ is the first pair that repeats itself in (23). Let the repeated pair be in the tth place (in accordance with what was established earlier we can regard $1 < t < m^2 + 1$), i.e. $\langle \bar{u}_t, \bar{u}_{t+1} \rangle = \langle 1, 1 \rangle$. This means that both u_t and u_{t+1}, when divided by m, give 1 as a remainder. It follows that their difference is exactly divisible by m. But

$$u_{t+1} - u_t = u_{t-1},$$

so that the $(t - 1)$th Fibonacci number is divisible by m.

We have thus proved the following theorem:

Theorem: *Whatever the whole number m, at least one number divisible by m can be found among the first m^2 Fibonacci numbers.*

Note that this theorem does not state anything about exactly which Fibonacci number will be divisible by m.

It only tells us that the first Fibonacci number divisible by m should not be particularly large.

III

FIBONACCI NUMBERS AND CONTINUED FRACTIONS

1. We consider the expression

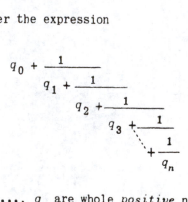

$$q_0 + \cfrac{1}{q_1 + \cfrac{1}{q_2 + \cfrac{1}{q_3 + \cfrac{1}{\cdots + \cfrac{1}{q_n}}}}} \tag{24}$$

where q_1, q_2, ..., q_n are whole *positive* numbers and q_0 is a whole *non-negative* number. Thus in contrast to the numbers q_1, q_2, ..., q_n, the number q_0 can equal zero. We shall keep this somewhat special position of the number q_0 in mind, and not mention it specially on each occasion.

The expression (24) is called a *continued fraction* and the numbers q_0, q_1, ..., q_n are called the *partial denominators* of this fraction.

Sometimes continued fractions are also known as chain fractions. They are of use in a wide assortment of mathematical problems. The reader who wants to study them in greater detail is referred to A.Ya. Khinchin, "Chain Fractions"*.

---

* Also to H.S. Wall, "Analytic Theory of Continued Fractions" (Van Nostrand) — Translator.

The process of transformation of a certain number into a continued fraction is called the *development* of this number into a continued fraction.

Let us see how we can find the partial denominators of such an expansion of the ordinary fraction $\frac{a}{b}$.

We consider the Euclidean algorithm, as applied to the numbers a and b.

$$a = bq_0 + r_1,$$

$$b = r_1q_1 + r_2,$$

$$r_1 = r_2q_2 + r_3,$$

$$\cdots \cdots \cdots \quad (25)$$

$$r_{n-2} = r_{n-1}q_{n-1} + r_n,$$

$$r_{n-1} = r_nq_n.$$

The first of these equations gives us

$$\frac{a}{b} = q_0 + \frac{r_1}{b} = q_0 + \frac{1}{\dfrac{b}{r_1}}.$$

But it follows from the second equation of set (25) that

$$\frac{b}{r_1} = q_1 + \frac{r_2}{r_1} = q_1 + \frac{1}{\dfrac{r_1}{r_2}},$$

so that

$$\frac{a}{b} = q_0 + \cfrac{1}{q_1 + \cfrac{1}{\cfrac{r_1}{r_2}}} \; .$$

From the third equation of (25) we deduce

$$\frac{r_1}{r_2} = q_2 + \frac{r_3}{r_2} = q_2 + \cfrac{1}{\cfrac{r_2}{r_3}}$$

and therefore

$$\frac{a}{b} = q_0 + \cfrac{1}{q_1 + \cfrac{1}{q_2 + \cfrac{1}{\cfrac{r_2}{r_3}}}} \; .$$

Continuing this process to the end (induction!) we arrive, as is seen easily, at the equation

$$\frac{a}{b} = q_0 + \cfrac{1}{q_1 + \cfrac{1}{q_2 + \cdots + \cfrac{1}{q_n}}} \; .$$

By the very sense of the Euclidean algorithm, $q_n > 1$. (If q_n were equal to unity then r_{n-1} would equal r_n and r_{n-2} would have been divisible by r_{n-1} exactly, i.e. the whole algorithm would have terminated one step earlier.) This means that in place of q_n we can consider the expression $(q_n - 1) + \frac{1}{1}$, i.e. consider (q_n-1) the last but one partial denominator, and 1 the last. Such a convention turns out to be convenient for what follows.

The Euclidean algorithm as applied to a given pair of

natural numbers a and b is realized in a completely
definite and unique way. The partial denominators of the
development of $\frac{a}{b}$ into a continuous fraction are also
defined in a unique way by the system of equations des-
cribing this algorithm. Any rational fraction $\frac{a}{b}$,
therefore, can be expanded into a continued fraction in
one and only one way.

<u>2.</u> Let

$$\omega = q_0 + \cfrac{1}{q_1 + \cfrac{1}{q_2 + \cdots + \cfrac{1}{q_n}}} \qquad (26)$$

be a certain continued fraction, and let us consider the
following numbers

$$q_0, \quad q_0 + \frac{1}{q_1}, \quad q_0 + \cfrac{1}{q_1 + \cfrac{1}{q_2}}, \quad \cdots$$

These numbers, written down in the form of ordinary
simple fractions

$$\frac{P_0}{Q_0} = \frac{q_0}{1} ,$$

$$\frac{P_1}{Q_1} = q_0 + \frac{1}{q_1} ,$$

$$\frac{P_2}{Q_2} = q_0 + \cfrac{1}{q_1 + \cfrac{1}{q_2}} ,$$

. . . .

$$\frac{P_n}{Q_n} = \omega,$$

are called *convergent* fractions of the continued fraction ω

Note that the transition from $\dfrac{P_k}{Q_k}$ to $\dfrac{P_{k+1}}{Q_{k+1}}$ is realized by the replacement of the last of those partial denominators which took part in the construction of this convergent fraction, i.e. q_k, by $q_k + \dfrac{1}{q_{k+1}}$.

3. The following lemma plays an important role in the theory of continued fractions.

Lemma: *For every continued fraction (26) the following relationships obtain:*

$$P_{k+1} = P_k q_{k+1} + P_{k-1}, \tag{27}$$

$$Q_{k+1} = Q_k q_{k+1} + Q_{k-1}, \tag{28}$$

$$P_{k+1}Q_k - P_k Q_{k+1} = (-1)^k. \tag{29}$$

We prove all these equations simultaneously by induction over k.

We shall prove them first for $k = 1$.

$$\frac{P_1}{Q_1} = q_0 + \frac{1}{q_1} = \frac{q_0 q_1 + 1}{q_1} .$$

Since the numbers $q_0 q_1 + 1$ and q_1 are prime to each other, the fraction $\dfrac{q_0 q_1 + 1}{q_1}$ is reduced to its lowest

terms. The fraction $\frac{P_1}{Q_1}$ is in its lowest terms according to the definition, and equal fractions in their
lowest terms have equal numerators and equal denominators.
This means that $P_1 = q_0 q_1 + 1$ and $Q_1 = q_1$.

$$\frac{P_2}{Q_2} = q_0 + \cfrac{1}{q_1 + \cfrac{1}{q_2}} = \frac{q_0(q_1 q_2 + 1) + q_2}{q_1 q_2 + 1} . \tag{30}$$

The greatest common divisor of the numbers $q_0(q_1 q_2 + 1) +$
$+ q_2$ and $q_1 q_2 + 1$ equals $(q_2, q_1 q_2 + 1)$ on the basis of
section 7 of II, and on the basis of the same proposition
it also equals $(q_2, 1)$, i.e. 1. This means that the
fraction on the right-hand side in (30) is in its lowest
terms, and therefore

$$P_2 = q_0(q_1 q_2 + 1) + q_2 = (q_0 q_1 + 1)q_2 + q_0 = P_1 q_2 + P_0$$

and

$$Q_2 = q_1 q_2 + 1 = Q_1 q_2 + Q_0.$$

The equation

$$P_2 Q_1 - P_1 Q_2 = (-1)^1$$

is easily verified.

The basis of the induction is thus proved.

Let us now suppose that the equations (27), (28) and
(29) are true and let us consider the convergent fraction

$$\frac{P_{k+1}}{Q_{k+1}} = \frac{P_k q_{k+1} + P_{k-1}}{Q_k q_{k+1} + Q_{k-1}}.$$

The transition from $\frac{P_{k+1}}{Q_{k+1}}$ to $\frac{P_{k+2}}{Q_{k+2}}$ according to the observation made above is brought about by the replacement

of q_{k+1} in the expression for $\dfrac{P_{k+1}}{Q_{k+1}}$ by $q_{k+1} + \dfrac{1}{q_{k+2}}$; since

q_{k+1} does not come into the expressions for P_k, Q_k, P_{k-1}, Q_{k-1}, then

$$\frac{P_{k+2}}{Q_{k+2}} = \frac{P_k\left(q_{k+1} + \dfrac{1}{q_{k+2}}\right) + P_{k-1}}{Q_k\left(q_{k+1} + \dfrac{1}{q_{k+2}}\right) + Q_{k-1}} ,$$

or, remembering the inductive assumptions in (27) and (28),

$$\frac{P_{k+2}}{Q_{k+2}} = \frac{P_{k+1}q_{k+2} + P_k}{Q_{k+1}q_{k+2} + Q_k} . \tag{31}$$

We now prove that the right-hand fraction in (31) is in its lowest terms. For this it is sufficient to prove that its numerator and denominator are mutually prime.

Let us suppose that the numbers $P_{k+1}q_{k+2} + P_k$ and $Q_{k+1}q_{k+2} + Q_k$ have a certain common divisor $d > 1$. The expression

$$(P_{k+1}q_{k+2} + P_k)Q_{k+1} - (Q_{k+1}q_{k+2} + Q_k)P_{k+1}$$

should then be divisible by d. But by the inductive assumption (29) this expression equals $(-1)^{k+1}$ and cannot be divided by d.

Thus the right-hand side of (31) is in its lowest terms, and (31) is therefore an equation between two fractions reduced to their lowest terms. This means that

$$P_{k+2} = P_{k+1}q_{k+2} + P_k$$

and that

$$Q_{k+2} = Q_{k+1}q_{k+2} + Q_k .$$

To complete the proof of the inductive transition it remains to show that

$$P_{k+2}Q_{k+1} - P_{k+1}Q_{k+2} = (-1)^{k+1} . \qquad (32)$$

But in view of what was proved above,

$$P_{k+2}Q_{k+1} - P_{k+1}Q_{k+2} =$$

$$= P_{k+1}q_{k+2}Q_{k+1} + P_k Q_{k+1} - P_{k+1}q_{k+2}Q_{k+1} - P_{k+1}Q_k,$$

and (32) follows directly from the inductive assumption (29). In this way the inductive transition is established and the whole lemma is proved.

Corollary.

$$\frac{P_{k+1}}{Q_{k+1}} - \frac{P_k}{Q_k} = \frac{(-1)^k}{Q_k Q_{k+1}} . \qquad (33)$$

Since partial denominators of continued fractions are positive whole numbers, it follows from the above lemma that:

$$P_0 < P_1 < P_2 < \cdots ,$$

$$Q_0 < Q_1 < Q_2 < \cdots \qquad (34)$$

This simple yet important observation will be made more exact later in the book.

4. We now apply the lemma of section 3 to describe all continued fractions with partial denominators equal to unity. For such fractions we have the following interesting theorem.

Theorem: *If a continued fraction has n partial denominators and each of these partial denominators equals unity, the fraction equals* $\dfrac{u_{n+1}}{u_n}$.

<u>Proof.</u> Let us denote the continued fraction with n unit partial denominators by a_n. Obviously

$$a_1, \ a_2, \ \cdots, a_n$$

are consecutive convergent fractions of a_n.

Let

$$a_k = \frac{P_k}{Q_k} .$$

As

$$a_1 = 1 = \frac{1}{1}$$

and

$$a_2 = 1 + \frac{1}{1} = \frac{2}{1} ,$$

therefore $P_1 = 1$, $P_2 = 2$. Further, $P_{n+1} = P_n q_{n+1} + P_{n-1} = P_n + P_{n-1}$. Therefore (compare I, section 8) $P_n = u_{n+1}$.

Similarly, $Q_1 = 1$, $Q_2 = 1$ and $Q_{n+1} = Q_n q_{n+1} + Q_{n-1} = Q_n + Q_{n-1}$, so that $Q_n = u_n$. This means that

$$a_n = \frac{u_{n+1}}{u_n} . \tag{35}$$

The reader should compare this result with formulae (12) and (29).

<u>5.</u> Suppose we are given two continued fractions ω and ω':

$$\omega = q_0 + \cfrac{1}{q_1 + \cfrac{1}{q_2 + \cdots}} , \qquad \omega' = q_0' + \cfrac{1}{q_1' + \cfrac{1}{q_2' + \cdots}}$$

while

$$q_0' \geqslant q_0, \quad q_1' \geqslant q_1, \quad q_2' \geqslant q_2, \quad \cdots \qquad (36)$$

Let us denote the convergent fractions of $\boldsymbol{\omega}$ by

$$\frac{P_0}{Q_0}, \frac{P_1}{Q_1}, \frac{P_2}{Q_2}, \quad \cdots$$

and the convergent fractions of $\boldsymbol{\omega'}$ by

$$\frac{P_0'}{Q_0'}, \frac{P_1'}{Q_1'}, \frac{P_2'}{Q_2'}, \quad \cdots$$

From the results of the lemma of section 3 it is easy to detect that in view of (36)

$$P_0' \geqslant P_0, \quad P_1' \geqslant P_1, \quad P_2' \geqslant P_2, \quad \cdots$$

and

$$Q_0' \geqslant Q_0, \quad Q_1' \geqslant Q_1, \quad Q_2' \geqslant Q_2, \quad \cdots$$

Obviously, the smallest value of any partial denominator is unity. This means that if all the partial denominators of a certain continued fraction are unity the numerators and denominators of its convergent fractions increase more slowly than those of the convergent fractions of any other continued fraction.

Let us estimate to what extent this increase is slowed down. Obviously, discounting the continued fractions whose partial denominators are unity, the slowest to increase are the numerators and denominators of the convergent fractions of that continued fraction one of whose partial denominators is 2 and the remaining ones unity. Such continued fractions are also connected with Fibonacci numbers as shown by the following lemma:

Lemma. *If the continued fraction $\boldsymbol{\omega}$ has as its partial denominators the numbers $q_0, q_1, q_2, \ldots, q_n$, while*

$$q_0 = q_1 = q_2 = \cdots = q_{i-1} = q_{i+1} = \cdots = q_n = 1, \ q_i = 2 \ (i \neq 0)$$

then

$$\omega = \frac{u_{i+1}u_{n-i+3} + u_i u_{n-i+1}}{u_i u_{n-i+3} + u_{i-1}u_{n-i+1}}.$$

Proof of this lemma is carried out by induction over i.   If $i = 1$, then for any n

$$\omega = 1 + \cfrac{1}{2 + \cfrac{1}{1 + \cfrac{\ddots}{\ \ + \cfrac{1}{1}}}}$$

$n - 1$ partial $\{$ denominator

or, in view of what was proved at the beginning of this section,

$$\omega = 1 + \cfrac{1}{2 + \cfrac{1}{\alpha_{n-1}}} = 1 + \cfrac{1}{2 + \cfrac{u_{n-1}}{u_n}} = 1 + \cfrac{1}{\cfrac{2u_n + u_{n-1}}{u_n}} =$$

$$= 1 + \frac{u_n}{u_{n+2}} = \frac{u_{n+2} + u_n}{u_{n+2}},$$

or, putting $u_0 = 0$

$$\omega = \frac{u_2 u_{n+2} + u_1 u_n}{u_1 u_{n+2} + u_0 u_n}$$

Thus the basis of induction has been proved.

Let us now suppose that for any n

i partial $\{$ denominators

$$1 + \cfrac{1}{1 + \cfrac{\ddots}{\ \ + 1 + \cfrac{1}{2 + \cfrac{1}{\alpha_{n-i}}}}} =$$

$$= \frac{u_{i+1}u_{n-i+3} + u_i u_{n-i+1}}{u_i u_{n-i+3} + u_{i-1}u_{n-i+1}} . \tag{37}$$

Let us take the continued fraction

$$\left. \begin{matrix} i + 1 \text{ partial} \\ \text{denominators} \end{matrix} \right\{ \quad 1 + \cfrac{1}{1 + \cdots} \cdots + 1 + \cfrac{1}{2 + \cfrac{1}{a_{n-i-1}}}$$

It can obviously be considered thus:

$$\left. \begin{matrix} i \text{ partial} \\ \text{denominators} \end{matrix} \right\{ \quad \overbrace{1 + 1}^{\cdots} \; \frac{}{1 + \cdots} \cdots + 1 + \cfrac{1}{2 + \cfrac{1}{a_{n-i-1}}} \tag{38}$$

The continued fraction below the dotted line in (38) is, by (37), equal to

$$\frac{u_{i+1}u_{n-i+2} + u_i u_{n-i}}{u_i u_{n-i+2} + u_{i-1}u_{n-i}} .$$

The whole fraction (38) therefore equals

$$1 + \cfrac{1}{\cfrac{u_{i+1}u_{n-i+2} + u_i u_{n-i}}{u_i u_{n-i+2} + u_{i-1}u_{n-i}}} = \frac{(u_i + u_{i+1})u_{n-i+2} + (u_{i-1} + u_i)u_{n-i}}{u_{i+1}u_{n-i+2} + u_i u_{n-i}} =$$

$$= \frac{u_{i+2}u_{n-i+2} + u_{i+1}u_{n-i}}{u_{i+1}u_{n-i+2} + u_i u_{n-i}} .$$

Thus the inductive transition has been proved and so has the whole lemma.

Corollary: *If not all the partial denominators of the contin-ued fraction ω are unity, $q_o \neq 0$, and there are no less than n of these partial denominators, then, on writing ω in the form* *of an ordinary fraction* $\dfrac{P}{Q}$, *we have*

$$P > u_{i+1}u_{n-i+3} + u_iu_{n-i+1} > u_{i+1}u_{n-i+2} + u_iu_{n-i+1} = u_{n+2},$$

and similarly,

$$Q > u_{n+1}.$$

A substantial role is of course played here by the lemma of section 3, on the basis of which we obtain only fractions in their lowest terms in the process of "contracting" a continued fraction into a vulgar one. Therefore no diminution of numerators and denominators of the fractions obtained due to "cancelling" will take place.

6. <u>Theorem</u>: *For a certain a the number of steps in the Euclidean algorithm applied to the numbers a and b equals n − 1 if $b = u_n$, and for any a it is less than n − 1 if $b < u_n$.*

<u>Proof:</u> The first part of the theorem can be proved quite simply. It is sufficient to take as a the Fibonacci number following b, i.e. u_{n+1}. Then

$$\frac{u_{n+1}}{u_n} = a_n.$$

The continued fraction a_n has n partial denominators, i.e. the number of steps of the Euclidean algorithm as applied to the numbers a and b equals $n - 1$.

To prove the second part of the theorem, we suppose the contrary, i.e. that the number of steps of this particular algorithm is not less than $n - 1$. Let us expand the ratio $\dfrac{a}{b}$ into the continued fraction ω. Obviously ω will have no less than n partial denominators (in fact one more than the length of the Euclidean algorithm). As b is not a Fibonacci number, not all the partial denominators of ω will be unity, and therefore, according to the corollary of the lemma in section 5, $b > u_n$, which contradicts the conditions of the theorem.

This theorem means that the Euclidean algorithm as applied to neighbouring Fibonacci numbers is in a sense the "longest".

<u>7.</u> We shall call the expression

$$q_0 + \cfrac{1}{q_1 + \cfrac{1}{q_2 + \cdots + \cfrac{1}{q_n + \cdots}}} \qquad (39)$$

an infinite continued fraction.

The definitions and results of the preceding sections can be extended quite naturally to infinite continued fractions.

Let

$$\frac{P_0}{Q_0}, \frac{P_1}{Q_1}, \ldots, \frac{P_n}{Q_n}, \ldots \qquad (40)$$

be a sequence (obviously an infinite one) of the convergent fractions of the fraction (39).

We shall show that this sequence has a limit.

With this aim in mind, we examine separately the sequences

$$\frac{P_0}{Q_0}, \frac{P_2}{Q_2}, \ldots, \frac{P_{2n}}{Q_{2n}}, \ldots \qquad (41)$$

and

$$\frac{P_1}{Q_1}, \frac{P_3}{Q_3}, \ldots, \frac{P_{2n+1}}{Q_{2n+1}}, \ldots \qquad (42)$$

From (33) and (34)

$$\frac{P_{2n+2}}{Q_{2n+2}} - \frac{P_{2n}}{Q_{2n}} = \frac{P_{2n+2}}{Q_{2n+2}} - \frac{P_{2n+1}}{Q_{2n+1}} + \frac{P_{2n+1}}{Q_{2n+1}} - \frac{P_{2n}}{Q_{2n}} =$$

$$= \frac{-1}{Q_{2n+2}Q_{2n+1}} + \frac{1}{Q_{2n+1}Q_{2n}} > 0$$

This means that the sequence (41) is an increasing one. In the same way, it follows from

$$\frac{P_{2n+3}}{Q_{2n+3}} - \frac{P_{2n+1}}{Q_{2n+1}} = \frac{1}{Q_{2n+3}Q_{2n+2}} - \frac{1}{Q_{2n+2}Q_{2n+1}} < 0$$

that the sequence (42) is a decreasing one.

Any term of the sequence (42) is greater than any term of the sequence (41). Indeed, let us examine the numbers

$$\frac{P_{2n}}{Q_{2n}} \text{ and } \frac{P_{2m+1}}{Q_{2m+1}}$$

and let us take the odd number k to be greater than $2n$ and $2m + 1$. It follows from (33) that

$$\frac{P_k}{Q_k} > \frac{P_{k+1}}{Q_{k+1}} , \tag{43}$$

and from the fact that (41) increases and (42) decreases it follows that

$$\frac{P_{k+1}}{Q_{k+1}} > \frac{P_{2n}}{Q_{2n}} \tag{44}$$

and

$$\frac{P_k}{Q_k} < \frac{P_{2m+1}}{Q_{2m+1}} . \tag{45}$$

Comparing (43), (44) and (45) we obtain

$$\frac{P_{2n}}{Q_{2n}} < \frac{P_{2m+1}}{Q_{2m+1}} .$$

From (33) and (34)

$$\frac{P_{n+1}}{Q_{n+1}} - \frac{P_n}{Q_n} = \frac{1}{Q_{n+1}Q_n} < \frac{1}{n^2},$$

and therefore, as n increases, the absolute value of the difference of $(n+1)$th and nth convergent fractions tends to zero.

From the above considerations it is possible to conclude that the sequences (41) and (42) have the same limit, which is also, obviously, the limit of (40). This limit is called the *value of the infinite continued fraction (39)*.

Let us prove now that any number can be the value of no more than one continued fraction. Let us take for this purpose two continued fractions ω and ω' (it does not matter whether they are finite or infinite).

Let q_0, q_1, q_2, ... and q_0', q_1', q_2', ... be their corresponding partial denominators. We shall show that it follows from the equation $\omega = \omega'$ that $q_0 = q_0'$, $q_1 = q_1'$, $q_2 = q_2'$ ... and so on.

Indeed, q_0 is the integral part of the number ω and q_0' is the integral part ω', so that $q_0 = q_0'$. Further, the continued fractions ω and ω' can be represented respectively in the form

$$q_0 + \frac{1}{\omega_1} \quad \text{and} \quad q_0' + \frac{1}{\omega_1'}$$

where ω_1 and ω_1' are again continued fractions. It follows from $\omega = \omega'$ and $q_0 = q_0'$ that $\omega_1 = \omega_1'$ also. This means that the integral parts of the numbers ω_1 and ω_1' are also equal, i.e. $q_1 = q_1'$. Continuing these arguments (induction!) we see that $q_2 = q_2'$, $q_3 = q_3'$, etc.

Since a rational number can always be expanded into a finite continued fraction, it follows from the foregoing proof that it cannot be expanded into an infinite continued fraction. It follows that the value of an infinite continued fraction must necessarily be an irrational number.

The theory of expansions of irrational numbers into continuous fractions represents a branch of theory of numbers which is rich in content and interesting in its results. We shall not delve deeply into this theory, but we shall consider only one example connected with Fibonacci numbers.

8. Let us find the value of the infinite continued fraction

$$1 + \cfrac{1}{1 + \cfrac{1}{1 + \dots}}$$

As we have proved above, this value is $\lim\limits_{n \to \infty} a_n$. Let us calculate this limit.

As has already been established in I, section 12, u_n is the nearest whole number to $\dfrac{a^n}{\sqrt{5}}$; this means that

$$u_n = \frac{a^n}{\sqrt{5}} + \theta_n,$$

where $\left| \theta_n \right| < \dfrac{1}{2}$, whatever n is.

Therefore, in view of the results of section 4,

$$\lim_{n \to \infty} a_n = \lim \frac{u_{n+1}}{u_n} = \lim_{n \to \infty} \frac{\dfrac{a^{n+1}}{\sqrt{5}} + \theta_{n+1}}{\dfrac{a^n}{\sqrt{5}} + \theta_n} =$$

$$= \lim_{n \to \infty} \frac{a + \dfrac{\theta_{n+1}\sqrt{5}}{a^n}}{1 + \dfrac{\theta_n \sqrt{5}}{a^n}} = \frac{\displaystyle\lim_{n \to \infty} \left(a + \dfrac{\theta_{n+1}\sqrt{5}}{a^n} \right)}{\displaystyle\lim_{n \to \infty} \left(1 + \dfrac{\theta_n\sqrt{5}}{a^n} \right)}.$$

But $\theta_{n+1}\sqrt{5}$ is a bounded quantity (its absolute value is less than 2) and a^n continues to increase indefinitely as n tends to infinity (because $a > 1$). This means

$$\lim_{n \to \infty} \frac{\theta_{n+1}\sqrt{5}}{a^n} = 0.$$

Also, for the same reasons,

$$\lim_{n \to \infty} \frac{\theta_n \sqrt{5}}{a^n} = 0,$$

and we obtain

$$\lim_{n \to \infty} a_n = a.$$

The theorem that has been proved means that the ratio of neighbouring Fibonacci numbers approaches a as their suffixes increase. This result can be used for the approximate calculation of the number a. (Compare the calculation of u_n in I, section 12.) This calculation produces a very small error, even when small Fibonacci numbers are taken. For example (correct to the fifth decimal place)

$$\frac{u_{10}}{u_9} = \frac{55}{34} = 1.6176,$$

and $a = 1.6180$. As we see, the error is less than 0.1%.

Of the errors involved in the approximate calculation of irrational numbers by convergent fractions it turns out that the number a represents the worst case. Any other number is describable by means of its convergent fractions in some sense more exactly than α. However, we shall not stop to consider this circumstance, interesting though it be.

IV

FIBONACCI NUMBERS AND GEOMETRY

1. Let us divide the unit segment AB into two parts in such a way (fig. 2) that the greater part is the mean proportional of the smaller part and the whole segment.

$$C_2 \qquad\qquad A \qquad C_1 \qquad\qquad B$$

Fig. 2.

For this purpose we denote the length of the greater part of the segment by x. Obviously, the length of the smaller part will be equal to $1 - x$ and the conditions of our problem give us the proportion:

$$\frac{1}{x} = \frac{x}{1-x}, \tag{46}$$

whence

$$x^2 = 1 - x. \tag{47}$$

The positive root of (47) is $\dfrac{-1 + \sqrt{5}}{2}$ so that the ratios in proportion (46) are equal to

$$\frac{1}{x} = \frac{2}{-1 + \sqrt{5}} = \frac{2(1 + \sqrt{5})}{(-1 + \sqrt{5})(1 + \sqrt{5})} = \frac{1 + \sqrt{5}}{2} = a$$

each. Such a division (at point C_1) is called *median section*. It is also called the *golden section*.

55

If the negative value of the root of the equation (47) is taken, the point of section C_2 lies outside the segment AB (this kind of division is called external section in geometry) as shown in fig. 2. It is easily shown that here, too, we are dealing with the golden section:

$$\frac{C_2B}{AB} = \frac{AB}{C_2A} = a.$$

2. The golden section appears quite frequently in geometry.

The side a_{10} of the regular decagon (fig. 3) inscribed in a circle of radius R is equal to

$$2R \sin \frac{360^{\circ}}{2.10},$$

i.e. it is $2R \sin 18^{\circ}$.

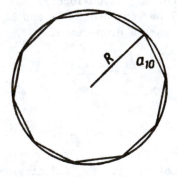

Fig. 3.

We now calculate $\sin 18^{\circ}$. From well known formulae of trigonometry we have

$$\sin 36^{\circ} = 2 \sin 18^{\circ} \cos 18^{\circ},$$

$$\cos 36^{\circ} = 1 - 2 \sin^2 18^{\circ},$$

so that

$$\sin 72^\circ = 4 \sin 18^\circ \times \cos 18^\circ (1 - 2 \sin^2 18^\circ). \qquad (*)$$

Since

$$\sin 72^\circ = \cos 18^\circ \neq 0,$$

then it follows from (*) that

$$1 = 4 \sin 18^\circ (1 - 2 \sin^2 18^\circ),$$

and therefore $\sin 18^\circ$ is one of the roots of the equation

$$1 = 4x(1 - 2x^2),$$

or

$$8x^3 - 4x + 1 = 0.$$

Factorizing the left-hand side of the latter equation we obtain

$$(2x - 1)(4x^2 + 2x - 1) = 0,$$

whence

$$x_1 = \frac{1}{2}, \quad x_2 = \frac{-1 + \sqrt{5}}{4}, \quad x_3 = \frac{-1 - \sqrt{5}}{4}.$$

As $\sin 18^\circ$ is a positive number, other than $\frac{1}{2}$, therefore $\sin 18^\circ = \frac{\sqrt{5} - 1}{4}$.

Thus

$$a_{10} = 2R \frac{\sqrt{5} - 1}{4} = R \frac{\sqrt{5} - 1}{2} = \frac{R}{\alpha}.$$

In other words, a_{10} equals the larger part of the radius of the circle, which has been divided by means of the golden section.

In practice, in calculating a_{10} we can use the ratio of neighbouring Fibonacci numbers (I, section 12 or III, section 8) instead of α and reckon approximately that

a_{10} is $\dfrac{8}{13} R$ or even $\dfrac{5}{8} R$.

3. Let us examine a regular pentagon. Its diagonals form a regular pentagonal star.

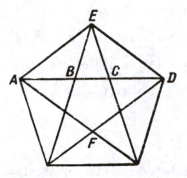

Fig. 4.

The angle AFD equals 108°, and the angle ADF equals 36°. Therefore, according to the sine rule

$$\frac{AD}{AF} = \frac{\sin 108^\circ}{\sin 36^\circ} = \frac{\sin 72^\circ}{\sin 36^\circ} = 2 \cos 36^\circ = 2\,\frac{1 + \sqrt{5}}{4} = \alpha.$$

Since it is obvious that $AF = AC$, then

$$\frac{AD}{AF} = \frac{AD}{AC} = \alpha,$$

and the segment AD is divided at C according to the golden section.

But from the definition of the golden section

$$\frac{AC}{CD} = \alpha.$$

Noting that $AB = CD$, we obtain

$$\frac{AC}{AB} = \frac{AB}{BC} = a.$$

Thus, of the segments

$$BC, \quad AB, \quad AC, \quad AD$$

each is a times greater than the preceding one.

It is left to the reader to prove that the equality

$$\frac{AD}{AE} = a$$

also holds.

4. Let us take a rectangle with sides a and b and let us proceed to inscribe in it the largest possible squares, as shown in fig. 5.

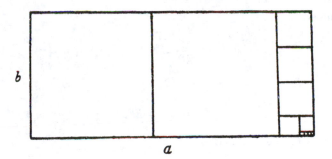

Fig. 5.

The arguments in II, section 2 show that such a process in the case of whole a and b corresponds to the Euclidean algorithm as applied to these numbers. The numbers of squares of equal size is in this case (III, section 1) equal to the corresponding partial denominators of the expansion of $\frac{a}{b}$ into a continued fraction.

If a rectangle whose sides are to each other as neighbouring Fibonacci numbers is divided into squares (fig.
6), then on the basis of III, section 4 all squares
except the two smallest ones are different.

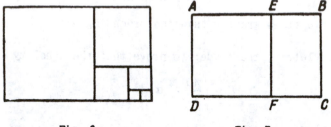

Fig. 6.   Fig. 7.

Now, let the ratio of the sides of a rectangle be equal to
a. (We shall call such rectangles "golden section rectangles" for short.) We now prove that after inscribing
the largest possible square into a golden section rectangle (fig. 7) we again obtain a golden section rectangle.

Indeed,

$$\frac{AB}{AD} = a,$$

also

$$AD = AE = EF,$$

since $AEFD$ is a square.

This means that

$$\frac{EF}{EB} = \frac{AB - EB}{EB} = a^2 - 1.$$

But $a^2 - 1 = a$, so that

$$\frac{EF}{EB} = a.$$

It is shown in fig. 8 how a golden section rectangle can be "nearly completely" exhausted by means of squares *I, II, III, ...* Each successive time a square is inscribed, the remaining figure is a golden section rectangle.

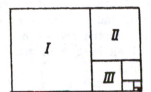

Fig. 8.

The reader should compare these arguments with sections 4 and 8 of the preceding chapter. We note that if a golden section rectangle *I* and squares *II* and *III* are inscribed in a square, as shown in fig. 9, the remaining rectangle turns out to be a golden section rectangle also. The proof of this is left to the reader.

5. Golden section rectangles seem "proportional" and are pleasant to look at. Things of this shape are convenient in use. Therefore, many "rectangular" objects of everyday use (books, matchboxes, suitcases and similar things) are given this particular form.

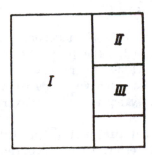

Fig. 9.

Various idealist philosophers of ancient and mediaeval
times raised the outward beauty of golden section rectangles and other figures which conform to the rules of
median section into an aesthetic and even a philosophic
principle. They tried to explain natural and social
phenomena in terms of the golden section and certain other
number relationships, and they carried out all kinds of
mystic "operations" on the number a and its convergent
fractions. It is clear that such "theories" have nothing
in common with science.

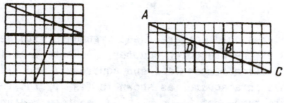

Fig. 10.   Fig. 11.

<u>6.</u> We shall round off our presentation with a little
geometrical joke. We shall demonstrate a "proof" that
64 = 65.

To do this we take a square of side 8 and we cut it up
into 4 parts as shown in fig. 10. We put the parts together to form a rectangle (fig. 11) of sides 13 and 5,
i.e. of area equal to 65.

The explanation of this phenomenon, puzzling at first
sight, is easily found. The point is that the points A,
B, C and D in fig. 11 do not really lie on the same
straight line, but are the vertices of a parallelogram,
whose area is exactly equal to the "extra" unit of area.

This plausible, but misleading "proof" of a statement
which is known beforehand to be incorrect (such "proofs"
are called sophisms) can be carried out even more "con-
vincingly" if we take a square of side equal to some

Fibonacci number with a sufficiently large even suffix,
u_{2n}, instead of a square with side 8. Let us cut up this
square into parts (fig. 12) and let us put these parts

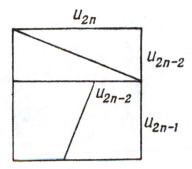

Fig. 12.

together to form a rectangle (fig. 13). The "empty space"
in the form of a parallelogram stretched along the
diagonal of the rectangle is equal in area to unity,

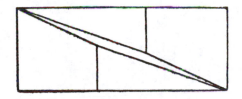

Fig. 13.

according to I, section 6. It is easily calculated that
the greatest width of this slit, i.e. the height of the
parallelogram, is equal to

$$\frac{1}{\sqrt{u_{2n}^2 + u_{2n-2}^2}}$$

If, therefore, we take a square with side 21 cm and
"convert" it into a rectangle with sides 34 cm and 13 cm,
the greatest width of the slit is found to be

$$\frac{1}{\sqrt{21^2 + 8^2}} \text{ cm,}$$

i.e. about 0.4 mm, which is difficult to detect by eye.

V

CONCLUSION

Not all the problems connected with Fibonacci numbers can be solved as easily as the ones we have considered. We shall indicate several problems the answers to which are either not known at all or can only be obtained by quite complicated means with the application of much more powerful methods of investigation.

1. Let u_n be divisible by a certain prime number p, while none of the Fibonacci numbers smaller than u_n is divisible by p. In this case we shall call the number p "the proper divisor of u_n". For example, 11 is the proper divisor of u_{10}, 17 is the proper divisor of u_9 and so on.

It turns out that any Fibonacci number except u_1, u_2, u_6 and u_{12} possesses at least one proper divisor.

2. The natural question arises: What is the suffix n of the Fibonacci number whose proper divisor is the given prime number p?

From II, section 13 we know that $n \leqslant p^2$. It is possible to prove that $n \leqslant p + 1$. Furthermore, it is possible to establish that if p is of the form $5t \pm 1$ then u_{p-1} is divisible by p, and if p is of the form $5t \pm 2$ then u_{p+1} is divisible by p. However, we have no formula to indicate the suffix of the term with the given proper divisor p.

3. We have proved in II, section 9 that all Fibonacci numbers with composite suffixes, except u_4, are composite themselves. The converse is not true, since, for example, $u_{19} = 4181 = 37 \times 113$. The question arises: is the number of all prime Fibonacci numbers finite or infinite, in other words, is there among all the prime Fibonacci numbers a greatest one? At this moment this question is still far from being solved.